FOALING
BROOD MARE AND FOAL
MANAGEMENT

FOALING
BROOD MARE AND FOAL MANAGEMENT

Ron and Val Males

HOWELL BOOK HOUSE INC.
230 Park Avenue, New York, N.Y. 10169

Distributed in the United States of America
and Canada by Howell Book House Inc.
230 Park Avenue, New York, N.Y. 10169

Published by Lansdowne Press, Sydney
a division of Kevin Weldon & Associates Pty Limited
372 Eastern Valley Way, Willoughby, NSW 2068, Australia
First published in the United States 1985
Reprinted 1986, 1989

© Ron and Val Males 1977
Produced in Australia by the Publisher
Typeset in Australia by G.T. Setters Pty Ltd, Sydney
Printed in Hong Kong by South China Printing Co.

ISBN 0 87605 851 9

Contents

Introduction

This book was compiled and written in response to constant requests for information about foaling.

Questions such as:

What should I expect?

How will I know if things are normal?

When should I call the Vet?

are asked daily by people who are venturing into a situation quite new to them. Their concern for the well-being of their mare through pregnancy, foaling and lactation is very real and justly so.

Others are often too embarrassed to ask questions as they feel they should somehow know the answers themselves because 'after all, foaling is a natural process!' These people often suffer endless pangs of fear, and even guilt, if something unforeseen happens.

Many excellent books about this same subject are presently available worldwide. Each has its individual merits and none is a substitute for another. Likewise, this book is not a substitute for others. Neither is it meant to be used as a replacement for experience or proper veterinary advice.

We agree with the old saying that a picture is worth a thousand words. Bearing this in mind we set out, hopefully, to show people by way of colour photographs and also explain to them what to expect in normal situations so they may perhaps be in a better position to judge when things are different to normal.

The black and white photographs should also be helpful in dispelling fear in unusual circumstances. Dozens of similar incidents are observed each year by breeders who care for large numbers of brood mares and so have the constant opportunity to observe, study, learn and, subsequently, to quickly evaluate the relevant situation.

Breeders of their first foals have rarely had this opportunity. It is to these people that this book is dedicated.

Ron & Val Males

RON & VAL MALES

A Normal Foaling

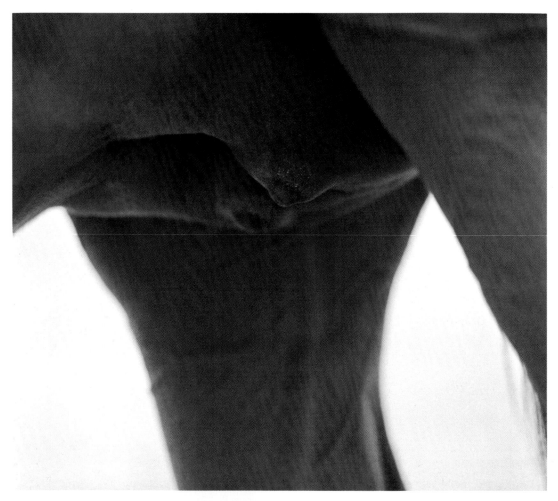

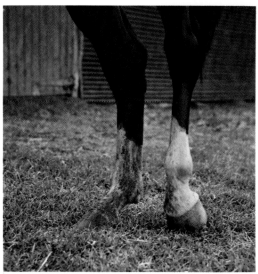

ABOVE: About three or four days prior to foaling, the udder will appear to be very large and full with a clear, grey wax-like coating all over it. Tiny white dots may surround the teats which will also appear full.

FAR RIGHT: Running milk.

LEFT: Accumulated milk and dirt have dried on the inside of the lower hind legs.

LEFT: The muscles along each side of the croup have relaxed and fallen in. The hair at the butt of the tail has been rubbed, giving it a ruffled appearance.

BELOW: At first glance the mare appears relaxed. . .but take a closer look. Her neck is high and rigid—the head has a worried expression—ears are held in a position indicative of stress.

BOTTOM: Another heavily pregnant mare stands in a relaxed manner. (For comparison to photo above only.) Note the low, stretched neck, peaceful expression, ears forward in an interested but unconcerned attitude.

TOP LEFT: The mare feels uncomfortable and stretches her loins, back and neck, yawning at the same time.

CENTRE LEFT: Automatically responding to a strong contraction, the mare swishes her tail and stamps a hind leg.

BOTTOM LEFT: Another strong contraction causes the mare to tuck up in the flank, lift her tail over her back and lean noticeably backwards in an effort to brace herself against the involuntary straining which is now beginning.

BELOW: The mare has moved to a different part of the paddock and, feeling colicky, prepares to lie down.

FAR LEFT: While still in a sitting position, the mare lifts her tail and commences straining. Fluid (from the ruptured placenta) runs out of the vagina as the inner membrane surrounding the foal make its first appearance.

ABOVE: Lying outstretched, with nose upturned, the mare continues straining. The first of the foal's forelegs is now visible.

LEFT: Sitting up again, she rests before the commencement of the next contraction. The hoof and pastern of the foal's foreleg are easily recognizable through the enveloping membrane (amnion).

LEFT: The tail lifts higher as another contraction pushes the foal's second foreleg into view.

BELOW: There is no need for concern when the mare gets to her feet at this stage of labour. In a matter of minutes she will lie down again.

ABOVE: Lying down once more, the mare begins to strain in earnest. The tail was moved aside only to photograph the foal's muzzle as it appeared.

TOP RIGHT: Another contraction pushes the foal further into the world. The force of this contraction was strong enough to push the foal's feet through the surrounding membrane. Look carefully at the white jelly-like substance protruding from the foal's hooves. This is protective padding which will break out very easily when the foal attempts to get to its feet.

BOTTOM RIGHT: The most difficult stage of labour is when the foal's shoulders are passing through the pelvis of the mare. This happens immediately the head is delivered.

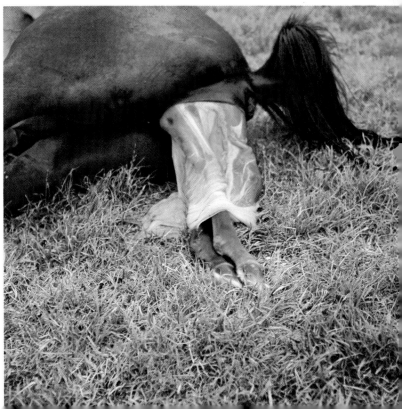

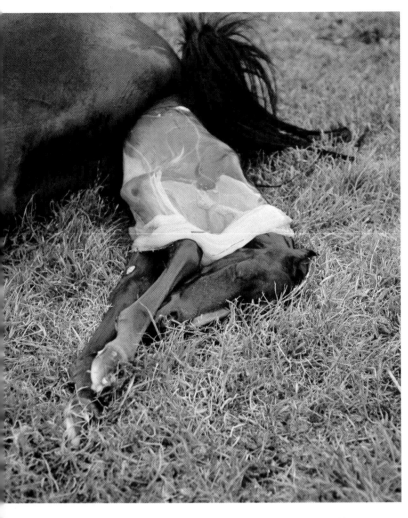

22

TOP LEFT: The major work finishes with delivery of the shoulders. A further contraction will expel the hindquarters.

BOTTOM LEFT: It's all over now and mum enjoys a well-earned rest! The hind legs remain in the mare's vagina until the foal moves about enough to dislodge them.

BELOW: The moment of recognition! The mare sits up and looks at her foal.

ABOVE: The foal is still attached to the placenta by the umbilical (navel) cord. This union should last as long as possible to allow complete transfer of blood from the placenta to the foal. If left alone, the cord will naturally begin to thin out just above its junction with the foal's navel. It will break of its own accord as the foal moves about or when the mare rises to her feet (whichever is the sooner).

RIGHT: Now standing, the mare carefully licks her foal as it lies on the ground. It is normal for the afterbirth to hang from the mare for some time before natural expulsion takes place.

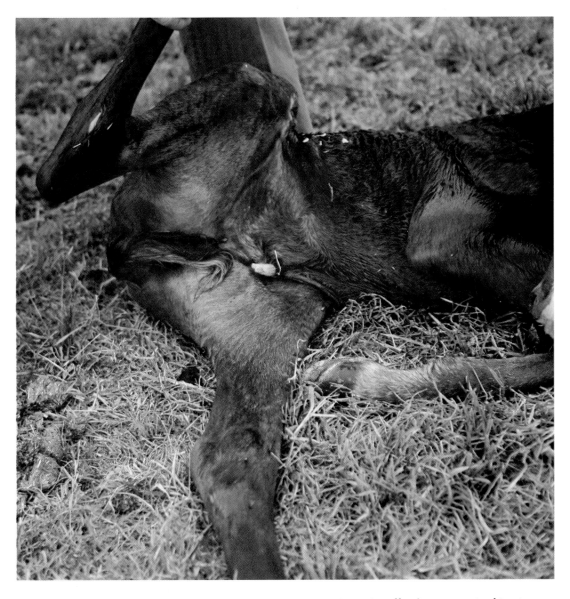

LEFT: Occasionally the new arrival is given a gentle nudge between licks!

ABOVE: The navel stump is about 4 cm (1½ in) long. It should wither and drop off in two to three weeks time.

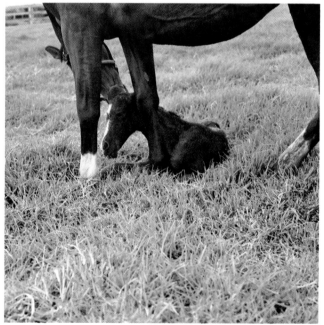

LEFT: The foal decides to try out her legs.

TOP: 'Oohh! What happened?'

ABOVE: Mum checks to make sure there is no damage.

TOP LEFT: *Foals get in the most awkward positions but seldom get hurt. Mares usually stand still (as this one does) until the foal moves away. Note the drawn up look in the mare's belly and flank areas as she experiences the first of several more contractions which will soon expel the afterbirth.*

BOTTOM LEFT: *If at first you don't succeed . . .*

ABOVE: *try, try, and try again!*

LEFT: Made it this time!

ABOVE: When first on its feet, the foal will stay in front of its mother keeping close to her breast and shoulders. When steady enough, it will gradually work its way along the mare's side to her flank before finding the teat sometime later.

ABOVE: *A new baby always attracts atten-*
tion, both from inquisitive mares...

RIGHT: *...and from children*

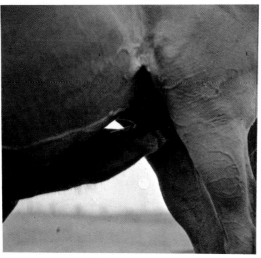

ABOVE: Newly born foals will often suck nearby objects such as fences, walls, fingers, etc. This one is sucking the grass.

LEFT: 'This is better than grass!'

RIGHT: The mare's udder after the foal has suckled. Relieved of pressure, the teats assume a flattish, shiny appearance.

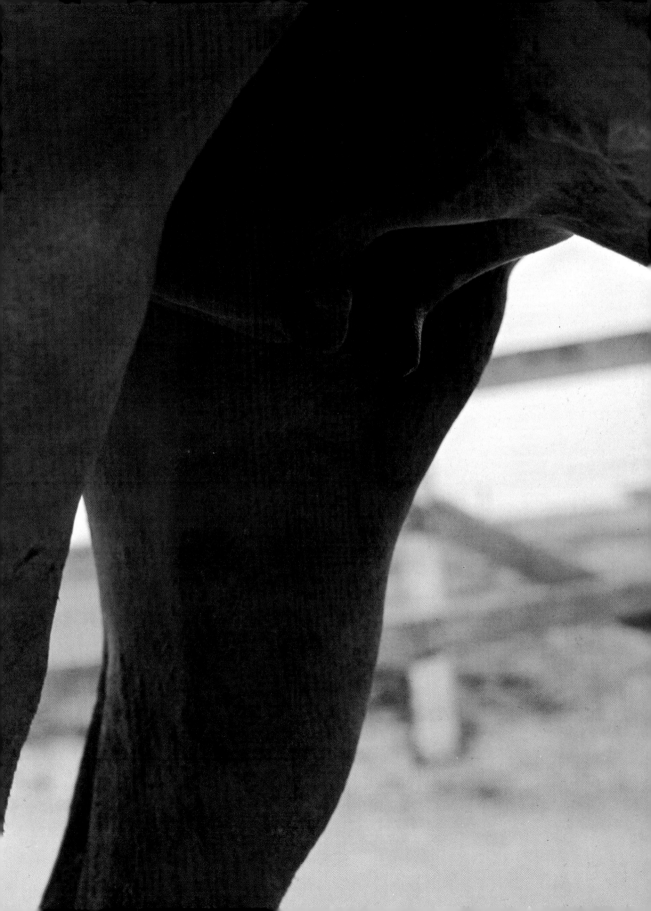

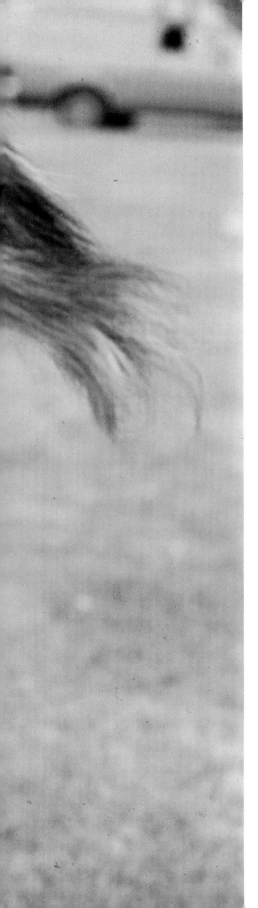

LEFT: The foal passes its first bowel motion which is called meconium. This very thick, dark, tar-like substance should be passed within a few hours of foaling.

BELOW: Meconium on the grass...a sure sign that the foal is clearing its bowel as nature intended.

BELOW CENTRE: Also as nature intended...the afterbirth has slipped away from the mare and lies the way it is usually found after natural expulsion.

BOTTOM: The umbilical cord is lying in the middle of the placenta; its jagged edges are normal. The mare and foal were not unduly disturbed before the cord ruptured and this allowed ample time for thinning out as the final blood was transferred to the foal. Less than 30 ml (about 1 fl oz) of blood drained from the cord when it broke.

RIGHT: This is where the foal was lying when the cord ruptured. Note the small amount of blood on the ground. Note also the little white pieces of padding which broke out of the foal's feet before it stood up.

CENTRE: The complete afterbirth (consisting of placenta and foetal membrane) has been spread out for easy inspection.

BOTTOM: The tips of both horns of the placenta should be rounded, smooth and intact.

FAR RIGHT: The fleshiest parts of a new foal are its buttocks. This area is commonly chosen for any early intramuscular injections which may be prescribed.

FAR LEFT: The mare's hindquarters, udder and tail should be washed as soon as practical after foaling. Cold water and soap or mild shampoo will effectively remove dried blood which will otherwise attract flies.

BELOW: Dried milk, blood and dirt which accumulate on the hind legs should also be removed when washing the mare.

BOTTOM: A soft but firm scrubbing brush should be used to loosen any stubborn or inground dirt.

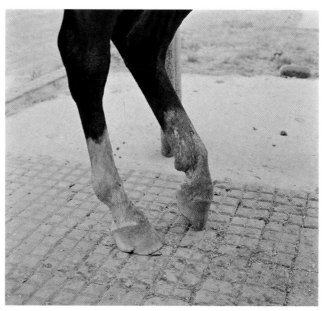

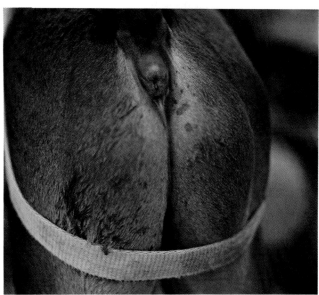

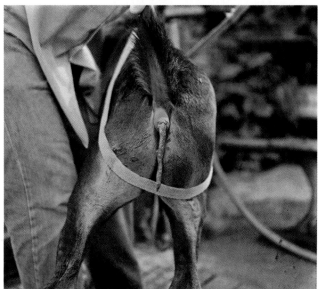

LEFT: The stud master takes this opportunity to trim the mare's feet. Note how the foal is being held and how quietly it is standing.

TOP: This is a different baby foal which has also passed its meconium. Note the tell-tale evidence of this by the way of the dark, wax-like small blob on the anus.

ABOVE: All is well! This soft mustard-coloured motion should follow soon after (a half to four hours) the baby foal passes its final meconium. It is a reliable guide that the foal is drinking and digesting milk. It also indicates that the foal is not constipated by meconium.

Less Usual Aspects of Foaling

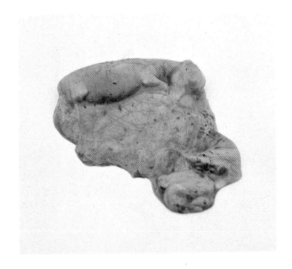

The following nineteen photographs
have been grouped together in a separate
section because they don't really belong
to the previous collection. Several of
these pictures depict everyday occurr-
ences which often go unnoticed whilst
others are extremely unusual and are
certainly not part of a normal foaling.
They should be of particular interest to
those who have not yet come across such
things in real life.

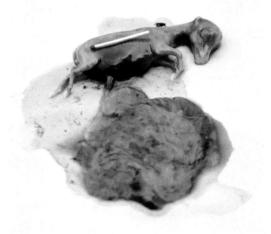

TOP: A foetus after its abortion at 61 days of
pregnancy. It is still enveloped in the amnion
and attached to the placenta.

CENTRE: The amnion has been removed
revealing the foetus which is clearly recog-
nizable as a miniature foal. Already its major
parts are well defined. Even the little hooves
are filled with their protective padding. (The
match and the fly provide excellent gauges
by which to judge the size of the foetus.)

BOTTOM: The umbilical cord is very
well developed as is the vulva which
indicates, even at this early stage, the sex of
the foetus as female.

FAR RIGHT: The tiny .tongue rests on the
match which has been placed in the mouth
above the lower gum.

PREVIOUS PAGE: What is better than one
healthy live foal? The answer of course is
two healthy live foals. Mrs Marty Stephens
from Henley Farm, Ebenezer, New South
Wales, proudly holds her pure Arabian mare
Royal Riff who, ten days earlier (and quite
unexpectedly) produced twin fillies. These
fillies (now two years old) are one of the very
few surviving sets of Arabian twins—
worldwide.

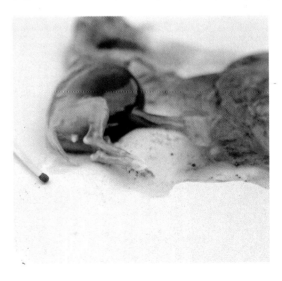

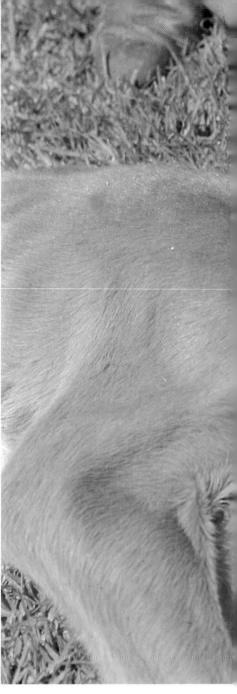

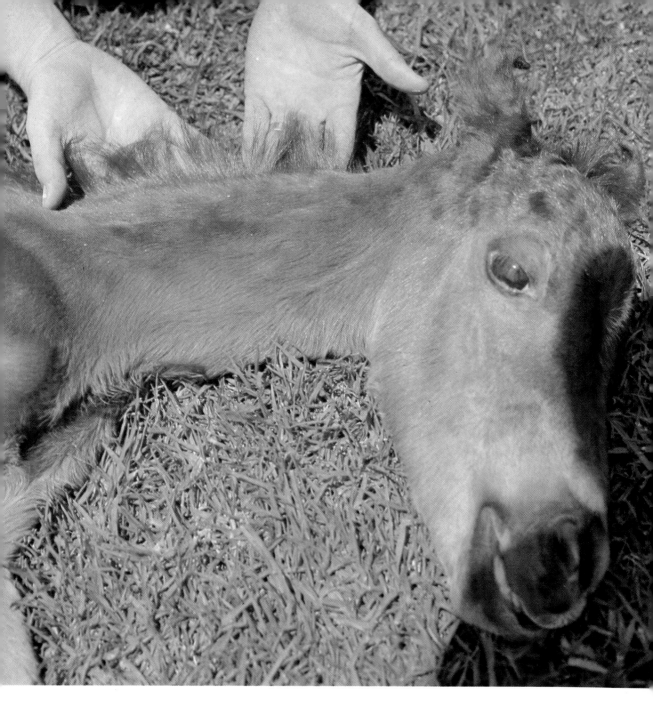

TOP LEFT: The mother of this colt had a pregnancy lasting 421 days (almost 14 months) before he was foaled. This colt was very small, weak and only survived 24 hours. The ruler on the foal measures 30 cm (about 12 in) by 60 cm (about 24 in); from this a fairly accurate estimate of size can be made. The hair in the brush of this foal's tail is as long as that of a normal 2½-month-old foal.

ABOVE: A further indication that this foal was carried well beyond the average period of pregnancy is the long hair in its mane.

LEFT: Milk teeth are already well developed.

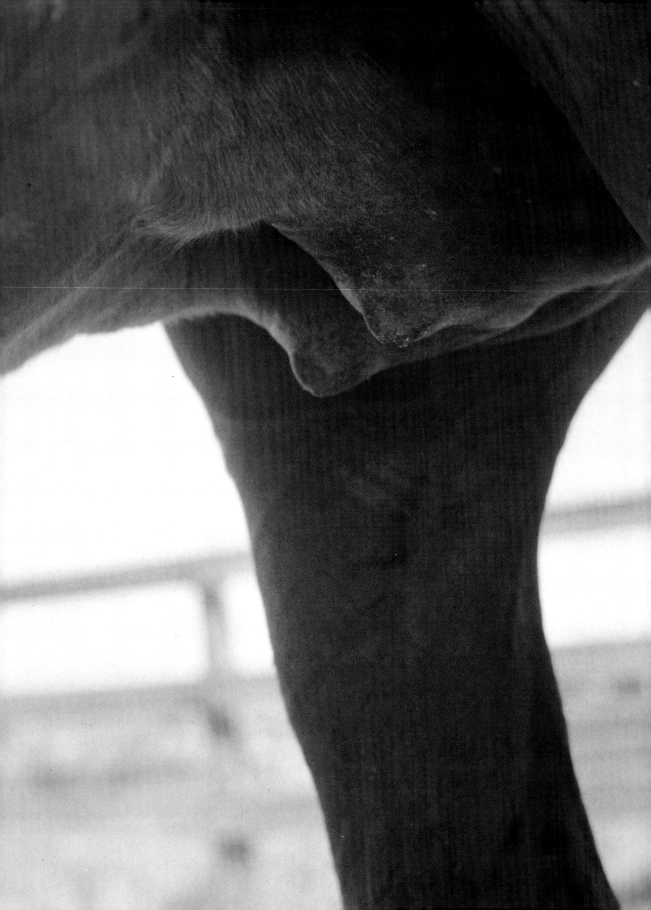

LEFT: The udder has almost reached its maximum development. The fine grey coating of wax is beginning to flake off around the teats. (This mare foaled ten hours later.)

BELOW: The udder of a mare that foaled without producing milk before the birth. It was necessary to put the mare in a stable and keep her confined until she was milking enough for the foal not to need a supplement.

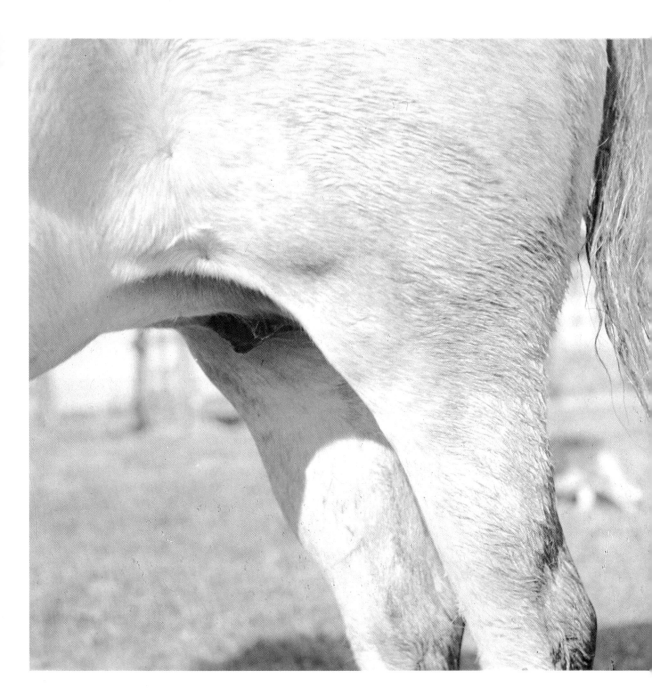

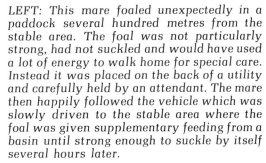

LEFT: This mare foaled unexpectedly in a paddock several hundred metres from the stable area. The foal was not particularly strong, had not suckled and would have used a lot of energy to walk home for special care. Instead it was placed on the back of a utility and carefully held by an attendant. The mare then happily followed the vehicle which was slowly driven to the stable area where the foal was given supplementary feeding from a basin until strong enough to suckle by itself several hours later.

TOP: The few silver white hairs above this newly born foal's eyebrow indicate that although it is presently dark chestnut, its coat will later turn grey in colour. The black skin around the eye lids and muzzle reinforce this indication.

ABOVE: This mare's natural fly protection works just as well for her son. During summer months when flies are prevalent the young foal will invariably stay close to its mother's hindquarters, thus her tail effectively swishes away the flies from the pair of them. (Photo: Keith Stevens.)

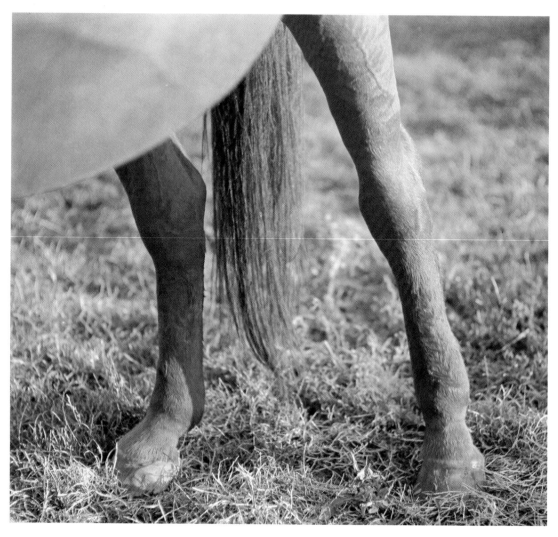

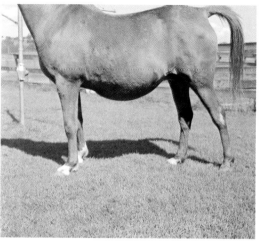

ABOVE: This mare has oedema (excessive accumulation of fluid in the tissues) of her legs as she nears foaling time. This commonly occurs during late pregnancy.

LEFT: An oedema extending from udder to brisket and, less noticeably, to lower limbs.

FAR RIGHT: Oedema is also evident around the vulva of this mare. Her tail is soiled from constant dribbling of soft manure which has also set up a crusty accumulation of debris between the hind legs from the vulva to the udder.

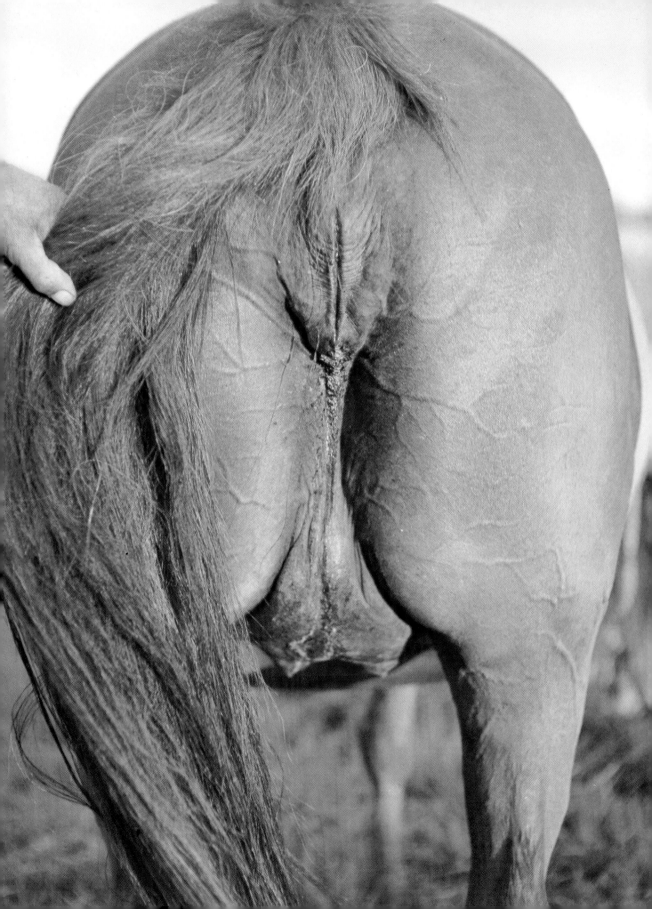

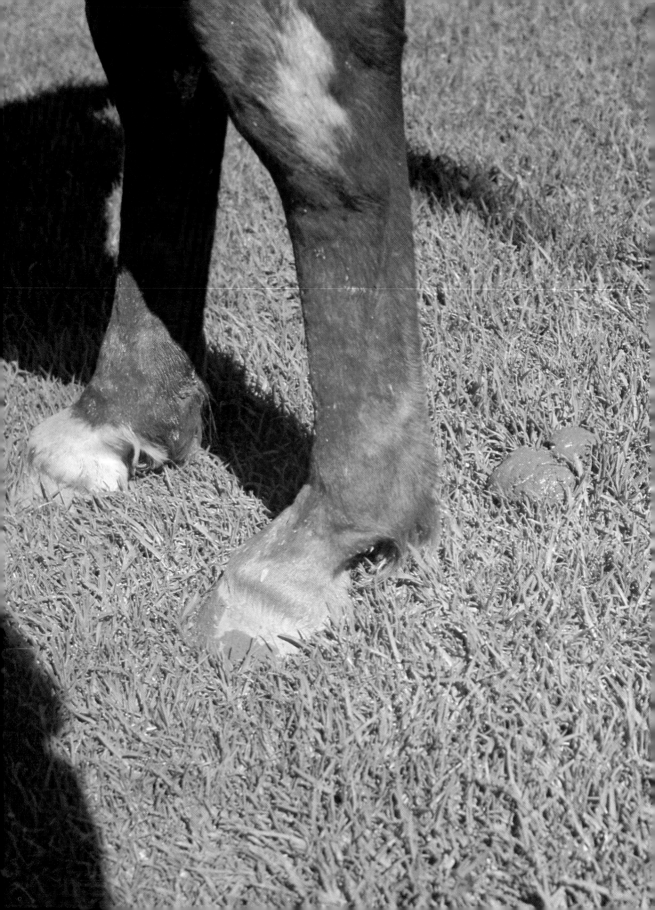

LEFT: When foaling time is close the mare will frequently pass small quantities of soft manure.

BELOW: Parasite control is an essential part of foaling hygiene. This manure was passed by a mare the day after foaling. Study it carefully. Many tiny red worms can be seen lying on its surface.

OVERLEAF: Sometimes nature overdoes things. This youngster was foaled with a large wart-like appendage to her upper lip. Although rather offputting at that time, the excess growth quickly dropped off without blemish a few days after an elastic band was tightly wound around its base.

Normal Pregnancy, Foaling and Foal Care

Large animals which have long-lasting pregnancies are not always punctual with delivery. They may safely give birth to healthy offspring some weeks before or after their due date.

So it is with mares—some foal on time but more often they are a few days early or late. Mares commonly foal three weeks earlier or later than expected. Most births within this range are successful and are accordingly considered 'normal'. Some foals happily arrive four weeks early or late. At times this period extends to five weeks either way.

Most people expect premature foals to be smaller and more delicate than those carried full term; few, however, realize that foals carried much beyond $52\frac{1}{2}$ weeks are also inclined to be smaller and more delicate than average. Like true premature foals, these little ones will need special care if they are to survive.

If it is suspected that foaling will take place outside the normal limits of pregnancy (less than 319 or more than 361 days) the mare should be watched carefully as her time approaches.

Heredity, environment, age, temperament and condition will each exert some influence on the ability of the mare to conceive, deliver and rear a healthy foal. With so many possible combinations of these and other factors, it is amazing that, overall, most mares manage to follow a fairly typical pattern of behaviour during pregnancy, foaling and lactation. Occasionally there will be 'way out' exceptions which are obviously not normal.

Others too will vary in behaviour—not only from the average but also from each other. Those, although different, may well be within the norm for a particular mare.

The following examples help to illustrate this point.

1. Annie, a ten-year-old, hard-fed mare in good condition, has foaled four weeks early every year since she was five years of age. She has produced and reared a healthy foal each time. Although unusual, the pattern she follows is normal for her.

2. Bessie, an eighteen-year-old, pasture-fed mare in excellent condition, foals three weeks late every year. Her foals too, have always been healthy. For her, that pattern of behaviour is normal.

3. Cassie, now twenty-two years of age, has throughout the past fifteen consecutive years foaled within a few days of her due date. Just as regularly (under varying conditions of management) she has produced skinny foals and very little milk. That too is her normal pattern of behaviour.

These examples also emphasize that each mare generally tends to follow her original 'tailor-made' pattern during subsequent pregnancies, so if a young mare produces her first foal three and a half weeks earlier than normal she will probably continue to follow that pattern for life. Any special arrangements for her future foalings should be made with this in mind.

Exceptions to the above rule are both the old and the over-fat mares, expecting their first foals. These mares are a law unto themselves: seldom do they conform to normal patterns during the latter part of pregnancy. Their second pregnancy is usually normal for them, and is the one they will tend to follow for the rest of their active breeding lives.

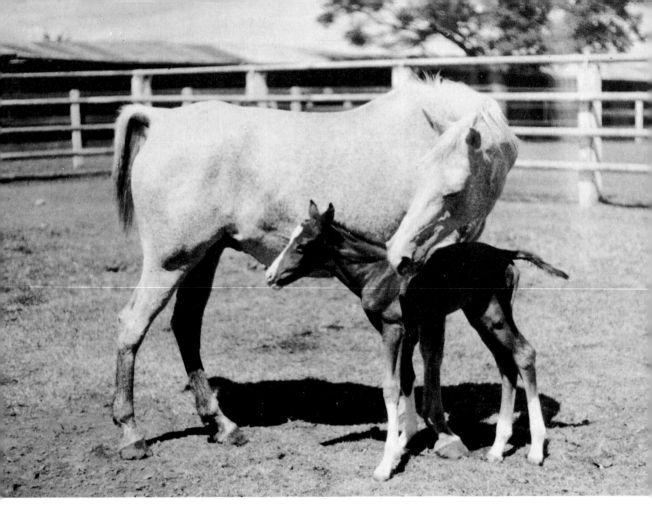

This foal was five and a half weeks premature. Note the lack of hair in mane and tail.

CONCEPTION

Pregnancy begins with conception. Conception takes place when an *ovum* (egg cell of the female) is fertilized by a *spermatozoon* (male sex cell, commonly called a sperm).

Normal conception results in the production of a single complete cell commonly called a *conceptus*. The conceptus develops by simple cellular division (i.e. splitting from 1 to 2, 4, 8, 16, and so on) until it assumes a vague outline and looks somewhat like a minute fluid-filled sac. At this stage its name changes to *embryo*. Later, when distinguishable as a tiny horse, it is referred to as a *foetus*. When the foetus is sufficiently developed for normal delivery and survival it becomes known as a foal.

NOTE: Though possible, it is unusual for conception to occur outside the mare's period of oestrus.

OESTRUS

Oestrus simply means receptivity. It is the name given to the periods when the mare is sexually responsive to the stallion.

Oestrus usually lasts for several days and occurs at regular intervals of approximately three weeks, during late spring, summer and autumn. It also occurs, but with less regularity during winter and early spring.

The actual word 'oestrus' is seldom used today, except in a technical sense. Stud masters prefer more practical terms. Some of these are as follows.

In season: presently used by a high proportion of breeders.

Horsing: an older term, once popular but seldom used today.

On heat: seldom used.

Heat period: frequently used as a substitute for 'oestrus period'.

Showing: an abbreviated term meaning showing signs of sexual desire ... often used with above terms.

On and Off: Commonly used when recording daily breeding details ... abbreviated terms meaning in and out of season respectively.

EXTERNAL SIGNS OF OESTRUS

Oestrus is easily recognized in the normal mare with access to a stallion. During her heat period the mare becomes very placid and behaves (more or less) in a typical manner.

- She will look at the stallion with an eager expression and ears forward,
- nicker or whinny to him,
- walk over to him,
- smell his head, giving a 'half-hearted' strike and/or squeal at the same time,
- turn her hindquarters towards his head,
- back up to him,
- lift her tail out behind her buttocks or up over her croup,
- stretch out and strain as if to urinate but only discharge varying quantities of clear, colourless mucous,
- 'wink'—the common term used by stud masters worldwide to describe eversion (turning inside-out) of the vulva. Clearly visible (when the mare lifts her tail), winking appears as a twitching of the vulva which then contracts, frequently exposing the clitoris,
- urinate frequently,
- she may continually repeat one or more of these actions.

In other words, the average mare in oestrus, when exposed to a stallion, shows unmistakeable signs of being ready for mating.

If there are no other horses in her vicinity the mare's sexual desires will rarely be stimulated to any obvious extent, thus her heat periods will probably pass unnoticed by all except the most observant. Oestrus can cause relaxation of the vulva and (sometimes) occasional intermittent minor discharges of clear colourless mucous from the vagina. Because these signs are so very slight, they are usually overlooked.

SERVICE OR MATING

In their natural free state, stallions and mares are friends not enemies. Unfortunately, many people have only witnessed a mating between a stallion and mare under unfavourable conditions. Because of this they go on believing that the resultant confusion and resistance is normal. Nothing is further from the truth. Resistance is an almost unmistakeable sign that there is something wrong.

Under normal circumstances, when a mare is ready for mating she will stand quietly and unrestrained before, during, and after service.

EARLY PREGNANCY (UP TO FOUR MONTHS)

When the normal mare becomes pregnant her oestrus periods cease. They do not recommence until she foals (except in the case of an abortion).

Temperament and/or other regular habits often alter quite markedly during pregnancy. Such changes are caused by the influence of various hormones which are present during this time. Thus the highly strung or energetic mare may suddenly become extremely placid; likewise the finicky eater might become ravenous; etc. These differences in behaviour sometimes become apparent within the first few weeks of conception.

Their importance, though inconclusive, should never be overlooked.

Another, but less obvious, change is seen in the shape of the mare's vulva which usually assumes a tight, dry or shrunken appearance during pregnancy.

The pregnant mare may become aggressive to the stallion or she may remain very friendly towards him. If friendly, she might even exhibit some signs that could cause us to think she was in oestrus. The mare may back up to the stallion and urinate in an intermittent squirting fashion. Her ears will be mobile and her tail swish regularly from side to side. The inexperienced (and frequently the experienced) breeder may well feel confused by this behaviour which closely resembles oestrus; however in such a situation the true placidity of oestrus is always absent. As a result, if the stallion is ardent and attempts service, he will quickly get a pair of heels in return for his unwanted advances.

It is interesting to note that the genuine 'paddock stallion' (one who continually lives out with his own band of mares) will always ignore the pregnant mare's overtures, no matter how persistent she is. No doubt he has learned the hard way that the percentage of pregnant mares which will accept service is almost nil!

Regardless of our experience or external signs which point to the probability of pregnancy we can never be *absolutely sure* the mare is pregnant until:
1. She is pregnancy tested (positive) by the veterinary surgeon;
2. She produces a foal.
These statements, although seemingly dogmatic, are not made lightly.

It is so easy to believe a mare is pregnant when she isn't. She may never have been pregnant or, she may have conceived and aborted a little later. Only proper pregnancy diagnosis by a qualified veterinary surgeon can give us information such as this.

Because of the time, money and labour involved (and personal disappointment when the mare proves empty), it is extremely important whenever possible to have the mare pregnancy tested well before fifty days of suspected pregnancy have elapsed.

Some vets prefer early tests, others later ones; this can only be decided by the experience of the veterinary surgeon, the type of test he/she selects, and obviously by the facilities available at the time.

When pregnancy diagnosis is positive the mare may continue her usual activities for several months as during this time the foetus is so small that its size can't possibly interfere with the mare's normal capacity for work. Nonetheless the mare should be treated with the respect she deserves and never overworked or overstressed as these factors play unknown roles in equine abortion.

During the latter part of early pregnancy the average mare will seem to radiate good health. This period may be likened to its human counterpart when it is often said that women glow with pregnancy.

MID PREGNANCY (FOUR TO EIGHT MONTHS)

Mid pregnancy is normally a quiet or uneventful period during which the mare continues to thrive.

Theoretically, we should be able to say that the belly of the pregnant mare increases in size from the third or fourth month through to foaling. This would make pregnancy easy to recognize by way of simple observation. Unfortunately it isn't so.

Size, in fact, is of little significance and may or may not increase noticeably as pregnancy progresses. Some mares have enormous bellies when empty, others foal with hardly a sign that they were pregnant.

Many people continue working their mares during mid pregnancy and this

Size of the pregnant mare's belly is of little significance. This mare foaled five weeks after being photographed.

*Thin is **not** healthy. (Photographer unknown)*

doesn't seem to do any harm as long as the mare is kept in healthy condition and is not subjected to sudden hard or unusual work.

It is probably during this time that most people begin to mistakenly 'feed the mare for two' with the result that she becomes over fat. Not until the seventh month of pregnancy does the developing foetus make many demands on the normal dry mare; even these are not significant. After this time the foetus increases more rapidly in size (though still slower than we generally imagine) until the final six to ten weeks when maximum foetal growth takes place.

The pregnant mare with a suckling foal will, at this time, require special care if loss of condition is to be avoided. Briefly, this mare *must* be well fed in order to provide her with the nourishment she needs to produce adequate milk for her suckling. In addition, she must maintain her own body weight.

It is agreed that over-fat mares (those with lumps of fat and enormous crests like stallions) are to be avoided. It is equally important not to go to the other extreme and think that thin is healthy and normal. This is not true.

Any healthy mare who is getting a sufficient quantity of good feed is able to maintain her own body weight and rear a foal while pregnant. It cannot be over emphasized that the normal healthy mare *does not lose condition* just because she is rearing a big strong foal. She loses it because the quantity and/or quality of the feed she is receiving is well below her needs.

The weight of both the mare and her suckling foal must be added together and this total weight used as a guide to calculate the amount of feed the mare needs.

It should be remembered too that even young suckling foals have healthy appetites. Provision must therefore be made for them to have plenty to eat in order to thrive. Foals are normally weaned between four and nine months, depending on circumstances. They should always be weaned before late pregnancy.

Nutritional needs of the former 'wet' brood mare diminish quickly after weaning and become similar to those of the 'dry' brood mare (one who has not been suckling a foal). In fact for all practical purposes this mare will now be referred to as a 'dry' brood mare.

The dry brood mare should always be well fed during mid and late pregnancy. She should not be allowed to get thin or over fat. Her rations need not vary much from those of the empty dry mare. Remember the old saying and act accordingly: 'The eye of the master is the condition of the horse.'

LATE PREGNANCY (EIGHTH MONTH TO DELIVERY)

From the eighth month onward the foetus grows and gains weight at an ever-increasing rate. It is during this period that final development and preparation for birth take place. Hair covers the body; eyelashes, mane and tail grow; while other organs such as skin and intestines, etc., also reach final stages of development. Vital organs and systems normally reach maturity between three hundred and three hundred and forty days of pregnancy.

Almost without exception, any pregnancy which terminates before three hundred days is unsuccessful. Quite often a foetus is delivered alive with a strong heart beat. It may even begin to breathe and thus lead us to believe there is hope of survival. However this is virtually impossible as foals with immature vital organs and systems cannot survive. To do so they would need highly developed artificial means of sustenance similar to those now used for premature human babies, such as humidicribs.

So be it.

NORMAL FOALING

Successful pregnancies have a duration of approximately three hundred and forty days. The mare's expected foaling date is calculated from her last date of service and may effectively be worked out very quickly by the following method.

- Check (don't guess) the last date of service, e.g. 1 December 1976.
- Add one year and seven days, i.e. 8 December 1977.
- Take off one calendar month: result is 8 November 1977.

The result is an approximate date for foaling.

About four to six weeks before her due date, the average mare usually begins to 'spring' or develop an udder. 'Spring' means to start or rise suddenly and describes this situation to a 'T'.

At first, the signs of springing may be slight with only the teats showing a small amount of thickening. This thickening may disappear and return in a day or so. When it reappears there may also be a small swelling in front of the teats. This swelling will gradually increase in size and extend to the teats which will in turn become enlarged. For a week or ten days, the swelling may continue to appear and disappear with monotonous regularity, particularly if the mare is young and springing begins early. Somehow it always seems to be there when we are alone and gone when we want to show it to somebody else!

Within the next ten days or so the size of the udder will slowly increase daily. During the early morning it will be larger than later in the day as exercise seems to diminish its size. This gradual development will continue until the mare is within approximately five or six days of foaling. At this time its size will not noticeably decrease through the daytime.

About three or four days prior to foaling, the udder will appear to be very large and full with a clear, grey wax-like coating all over it. Tiny white dots may surround the teats, which will also appear full.

Within twelve to thirty-six hours of foaling the wax-like coating will begin to flake and peel off the udder which will then become very black and shiny. (This coating is so thin that many people never notice it.)

At approximately the same time the teats will become pointed and the mare may begin to show signs of running milk. At first there might only be a single drop of what looks like white candle wax, on the end of one or both teats. This usually drops off and is replaced sometime later (between a few seconds and an hour) by another drop. As time goes by, the drops will immediately be replaced as they drop off. Sometimes these drops become quite hard and will accumulate so they actually look like small white wax 'birthday cake' candles clinging to the end of the teats. (In some cases they reach a length of 5 cm [2 in].)

As pressure builds up inside the udder, the wax-like substance (which is actually dried colostrum) will stop forming and may be replaced by intermittent or continual streams of milk dripping from the teats down on to the inside of the hind legs and feet of the mare.

When the mare begins to run milk the muscles along her croup will relax and she will appear to have suddenly lost condition in that area. Not all mares run milk and some don't even drip; however the relaxation (or falling in) of the croup muscles always takes place and is a general guide that foaling will occur within the next twenty-four hours or so. If the mare is very fat the falling in of the croup muscles may be hard to see. Nevertheless it does take place and if the croup area is patted its muscles will feel soft and flabby.

At this juncture the mare can be expected to show signs of restlessness consistent with onset of labour.

Careful, regular observation (at least two-hourly) of the mare will now be necessary if she is to be watched while foaling. Mares have an uncanny knack of foaling when least expected.

The first stage of labour is the period when final preparations are made for delivery of the foal. In the average mare it extends from two to eight hours but in some instances might only last for a very brief period before delivery; alternatively it could extend to twenty-four hours without causing any problems (except to the impatient owner).

During the first stage of labour regular, wavelike, contractions extend along the uterine wall towards the cervix (the neck of the uterus, a muscular structure which when closed separates the uterus from the vagina). The cervix (which was so tightly closed through pregnancy) has to open wide enough during the first stage of labour to allow safe passage of the foal. Each uterine contraction gradually forces the foal towards the cervix which then slowly but surely opens in response to the pressure.

As the second stage of labour approaches the mare will become very agitated. As her time draws near she may frequently:
- stop eating,
- pass small quantities of very soft manure,
- urinate or strain (and relax the vulva),
- rub her tail and croup against a fence, wall or tree etc.,
- walk or run in an agitated manner,
- paw the ground,
- smell the ground as if to lie down,
- stretch her back, loins and neck,
- hollow or arch her back,
- yawn, . . .
- curl her top lip upwards,
- blow her nose,
- swish her tail,
- stamp alternate hind legs,
- tuck up in the belly and flank areas,
- eat in a seemingly 'compulsive' manner,
- lie down, or
- roll . . . gently or more actively . . . sometimes rubbing her neck and head along the ground.

Progressive development of the udder. (Taken at weekly intervals from six weeks prior to foaling.)

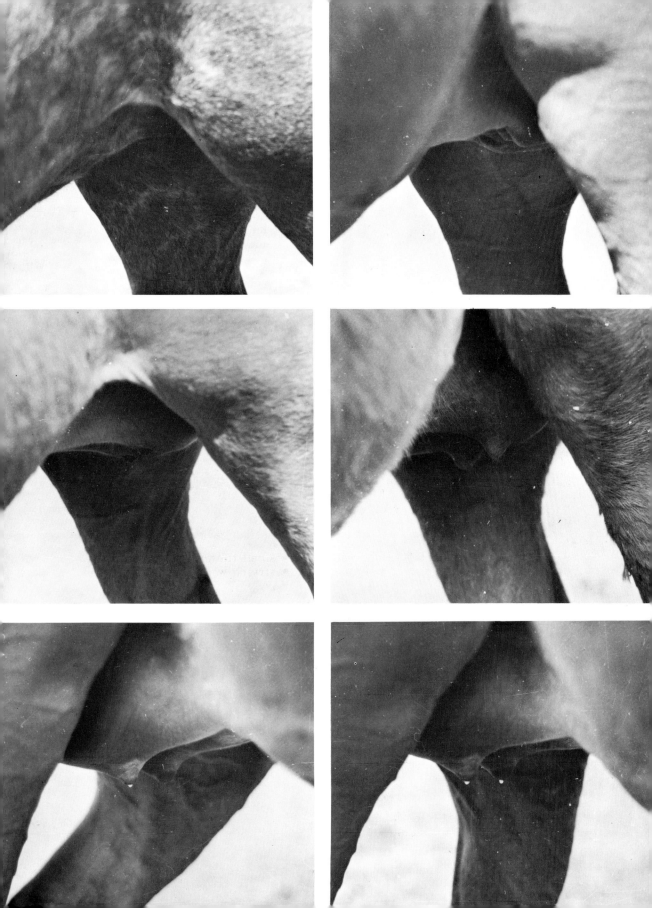

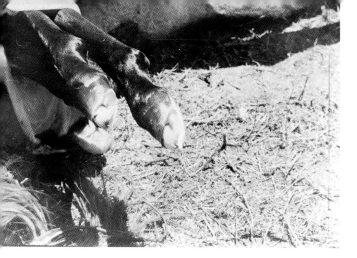

The foal's hooves are filled with a protective jelly-like substance.

The mare's legs stiffen with a severe contraction which eventually expels the head.

The first stage of labour finishes when the cervix is fully dilated.

Soon afterwards the second stage of labour begins, with the characteristic rupturing of the placenta. This sometimes happens when the mare is standing in a straining position. Other times it occurs when the mare is lying down. (Rupture of the placenta is often referred to simply as 'breaking of the waters'.) The fluid which subsequently gushes from the vagina is only a little darker in colour than normal urine. Because of this it is likely to be mistaken for urine, particularly if the mare is standing outstretched when her water breaks.

The remainder of the second stage of labour deals with the physical expulsion of the foal. When it begins, the mare will nearly always respond to its very strong regular contractions by lying down. Occasionally a mare will foal in a standing position but this is uncommon.

The most valuable pieces of equipment an attendant can have at this time are a **timepiece**, a **pad**, and a reliable **pen or pencil.**

The biggest help he/she can give the mare at this stage is to carefully *record* the actual *times* as each new phase occurs. For example:

1 a.m.: Placenta appeared to rupture—mare lay down.

1.05 a.m.: Membranes visible, etc.

This will prove invaluable if veterinary assistance is required (extremely rare). Most newcomers tend to panic in a foaling situation that doesn't exactly go by the book. As a result, when summoning help, they seldom remember the things which are most important, i.e. what happened and when it happened. NOTE: Contrary to general opinion, mares *do not* mind human company during foaling. In fact at this time they seem to enjoy company and if people are present the average mare will usually go to them for 'sympathy'.

Let us now assume that the placenta has ruptured and the mare is sitting down. From this point the average mare delivers her foal within twenty minutes but this can seem like forever if not accurately timed.

The first tangible sign of foaling is the appearance of the amnion. This is the shiny, inner foetal membrane which completely envelops the foal. It too contains a quantity of clear yellowish fluid which may also be mistaken for urine.

The mare then lies outstretched and will begin to strain with each regular contraction. Before long the first foreleg of the foal will appear. It should be slightly in advance of the second one which will soon follow.

The soles of the foal's feet will

This funny little liver-like object is called a **hippomane.** It is apparently formed by salts etc., which accumulate in the fluid within the placenta before foaling. (Commonly found on the ground after delivery of a foal.)

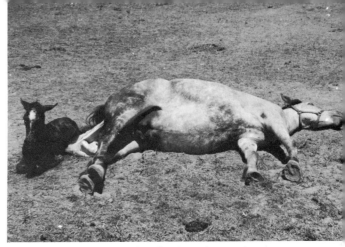

This mare is not dead. She is only resting after foaling.

eventually turn to face down towards the mare's hocks. This indicates correct positioning for delivery, i.e. 'spine to spine' with the mare. Prior to delivery the foal lies on its back in the uterus and only turns over into position for delivery as it begins to make its exit into the world.

Mares commonly frighten their attendants at this stage by rolling vigorously or by getting up and down several times. Both actions are normal and somehow seem to be associated with correct positioning of the foal.

The mare will continue to have very strong contractions. Each contraction causes more straining and further expulsion of the foal. After both forelegs are presented, it takes only a few minutes for the muzzle to come into view. The foal's muzzle should rest downwards on the forelegs just below its knees.

The mare may seem to experience some difficulty during/after delivery of the foal's head but it is seldom the head alone that causes this problem. The shoulders of the foal are comparatively wide so subsequently they are a tight fit as they pass through the pelvis.

At this stage the mare has to work very hard (i.e. 'labour') and may take advantage of several rest periods between contractions. She may even choose to get up and walk a few paces

before lying down again.

Nevertheless it should not be too long before a series of extremely strong contractions enable the mare to expel the foal's shoulders. The shoulders will quickly be followed by the back, loins and hindquarters.

After the hindquarters arrive it is usual for the mare to rest, outstretched, for a period ranging, in time from approximately a few minutes to a quarter of an hour. She may lie almost completely motionless (as if dead!) and during this time the foal's legs will remain in the vagina until dislodged by movement of the mare or the foal. Likewise the umbilical cord will remain attached to the foal until a sudden movement causes it to break.

Sometimes the foal is delivered, complete, in the amnion which will rupture easily as the foal moves on the ground. Other times this membrane breaks when the foal's forelegs first come into view. Either way is normal. If by any strange chance (rare) the amnion hasn't ruptured soon after delivery it may easily be torn with the fingers to uncover the foal's head.

Some people like to drag the foal around to the mare's head soon after delivery so the mare can see and lick her new baby. This shouldn't be done until the umbilical cord has ruptured of its

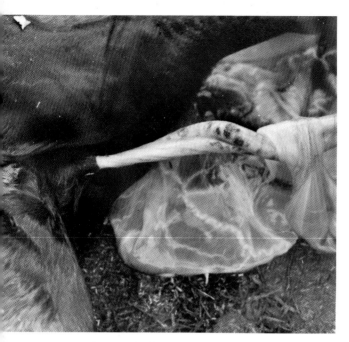

The cord thins near its junction with the foal's navel.

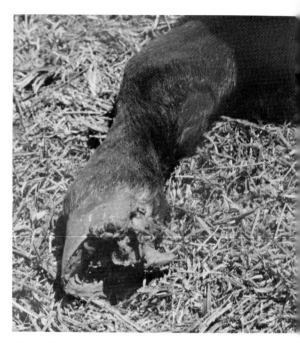

The jelly-like substance breaks out of the hooves when the foal moves about.

own accord. The longer the foal is attached to the placenta (or afterbirth) by this cord, the better. This allows final transfer of blood from the placenta to the foal. If left alone the cord will naturally begin to thin out just above its junction with the foal's navel. It will break when the foal moves about or when the mare gets to her feet.

The mare will eventually sit up and look at her foal before she stands up. When she does stand she may appear groggy and wander off away from the foal before returning to lick it. Some mares never lick their foals. Foals are clean when born and licking therefore does not 'clean them up', as some people imagine. However it can be very stimulating for a foal and no doubt helps to dry it.

More importantly, licking, nudging and gentle biting all play a major part in the successful mothering between mare and foal. This bond is essential for the well-being of the foal and should take place as soon as conveniently possible

after foaling (i.e. after the cord ruptures and the mare has rested).

It happens sometimes, that the foal gets up before its mother and is 'stolen' by another horse. For this reason the foaling mare should be separated from all except other foaling mares. Even then there is a slight risk of 'theft' which can only be prevented by careful observation.

When her foal wanders or is taken away from the mare before she mothers it the mare will often react by mothering either the ground where the foal was born, or the afterbirth when it is expelled. It then will take much time and perseverance to persuade the mare to accept her foal when it is returned to her.

When the foal first attempts to stand it usually makes a lot of unsuccessful efforts before it manages to get to its feet. During this time it can roll an incredible distance from where it was born.

If the mare foals on top of a steep hill for instance, the foal could quite easily

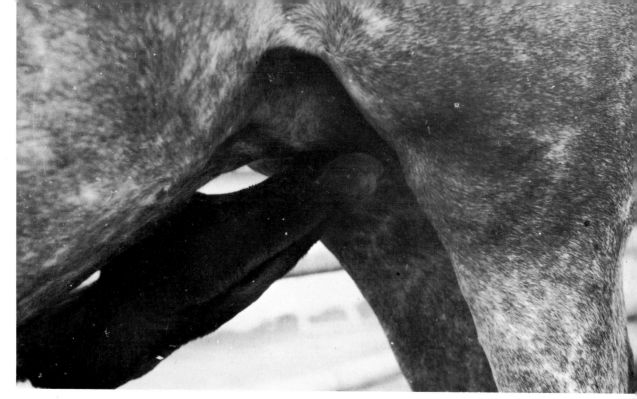
The new foal suckles the mare.

roll to the bottom and under a nearby fence. This is a very common occurrence which causes many early foal deaths. Likewise the foal could roll down a steep (or slight) bank into a creek or dam. If indoors it could become cast against a wall or jammed under the lower part of a door.

Careful observation again is the criterion during this critical part of the foal's life and can never be stressed too much. There is no room for complacency if foals are to be reared successfully.

After the foal is on its feet it should soon begin to seek a drink from its mother. At first the foal appears wobbly and will stand close by the mother's breast until its legs become steadier and more controllable. Within the hour it will turn to face the mare's tail and gradually fumble its way along her side to her udder.

At approximately the same time as the foal seeks its first drink, the third stage of labour, i.e. expulsion of the placenta, usually takes place. Under normal conditions the foal's sucking stimulates the uterus into a further series of contractions which should expel the placenta. Occasionally the mare will lose the placenta before getting to her feet after foaling; often it will be expelled while she is licking the foal as it lies on the ground.

If the afterbirth (as the placenta is often called) is not expelled naturally by the mare within a few hours of delivery, veterinary advice should be sought. It should never be removed by force.

Alternatively the vet may be asked for advice (before foaling) re possible retention of the afterbirth. It is hard to give hard and fast advice for this situation as most vets prefer to differ on the subject. In the absence of professional advice to the contrary, it is usual to seek veterinary assistance if the placenta is retained for longer than three or four hours. This will allow plenty of scope in cases where the vet has a long distance to travel or when he/she cannot attend for several hours.

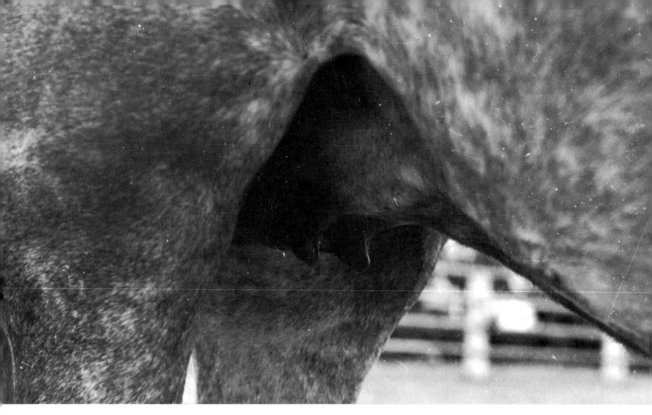

The udder after suckling. Note the shiny flattish appearance of the teats.

Putrification is followed closely by infection and can set in very quickly after foaling if the placenta is retained for too long in hot weather.

IMMEDIATE POST-NATAL CARE (FIRST 36 HOURS)

Owners are often so enthralled by their new foal's sex, colour and markings that essential details are overlooked.

The situation should be approached in a positive manner. First check to see that all is normal.

- Is the mare all right?
- Has the foal got two eyes, two ears, four functional legs, etc?
- Has the foal suckled?
- Has the foal passed its meconium?

A negative answer and/or neglect, to any of the above, could well be the first step towards the illness or death of the mare and/or foal.

Post-Natal Check List for Mare

The mare should be examined for signs of bruising, swelling and/or tearing of the vulva and (outer) vagina. Excessive swelling usually follows birth of a very large foal or it can indicate that labour has been longer and/or more difficult than usual.

Sometimes in the very rare case of mal-presentation, the foal's feet may tear through the vagina into the rectum. This tearing is usually repairable but it *is* serious and deserves immediate attention by a veterinary surgeon. If neglected it will almost certainly lead to infection and infertility. Because the mare's tail covers the region where this type of injury occurs it is often overlooked until it becomes infected and smelly.

Uncomplicated swelling of the vulva will subside within a few days to a week but because it is sometimes accompanied by vaginal bruising it is advisable to have the mare examined by a veterinarian if the swelling persists longer than twelve to twenty-four hours after foaling. Uncomplicated vaginal bruising will also heal and disappear spontaneously but it may take several weeks to do so. During this time it would be unwise to have the mare remated.

The mare should be checked for abnormal vaginal discharges. It is normal for her to discharge small quantities of blood-stained serum. Continual heavy bleeding (constant or intermittent) is not normal and should be treated seriously.

After the birth of a mare's first foal (or a subsequent large foal) the sphincter muscle which normally prevents involuntary discharge of urine occasionally becomes stretched. When this occurs the mare will pass urine frequently and uncontrolledly as she walks or trots. During this time it is advisable to wash the mare's hindquarters, tail and udder at least once each day. The sphincter should gradually regain most of its former elasticity and begin to function normally within a week.

As soon as practical after foaling, the mare's udder, hindquarters, tail and hind legs should be thoroughly washed with soap (or mild shampoo) and water. This will remove all the blood, milk and other accumulated matter which so readily attracts flies and invites bacterial infection. Disinfectants are not necessary and if applied may burn the udder, vulva or even the baby foal's lips (if it nuzzles the treated area). Nowadays most people with horses have a special washing area with running water and this makes the job very easy.

However, if these facilities are not readily available and the mare is quiet, there is no reason why she shouldn't be washed in the paddock or stable. In such an instance an extra person will probably be needed to hold the baby foal in front of the mare otherwise the foal will possibly run around and thus upset its mother.

Apart from the obvious reason of hygiene, washing the mare also serves to make her feel clean and thus more comfortable. Washing her has another advantage. It necessitates giving extra attention to the mare and therefore offers opportunity for further observation. Unusual or abnormal conditions which may otherwise remain undetected will almost certainly be noticed when mare and foal are at such close quarters.

The mare's udder should be examined to see that both teats are milking. For the first day or so the udder is likely to be very swollen. This may appear to be exaggerated if the foal only sucks from one teat instead of two; this doesn't really matter though as eventually it will learn to drink from both teats.

While the mare's udder is swollen it will also be very tender. Constant sucking by the foal (particularly if it has teeth) will often make the teats very sore. Sometimes they will even peel and bleed. Naturally this hurts the mare who will react by lifting a hind leg and/or squealing. This behaviour will cease as soon as the teats heal and toughen. It should not be misinterpreted as a sign that the mare is rejecting her foal.

During the first two weeks after foaling the mare will also tuck up in the belly and flank areas when the foal is sucking. Such tucking up is usually due to contractions of the uterus which occur spontaneously with sucking and cause varying degrees of discomfort.

Post-Natal Check List for Foal

The normal baby foal should appear reasonably strong, active and inquisitive. It should be on its feet within an hour or so of delivery and should suckle often (at least half hourly) and sleep just as regularly.

Mares produce a lot of milk during each twenty-four hours but, because they don't have a big storage capacity (like cows do), the foal drinks more often than most new foal owners ever imagine is normal. Likewise, the little foal's stomach has a small capacity (when compared to a calf) and can only comfortably hold a small quantity of milk at any one time.

These simple commodities could save a foal's life.

Within an hour or two of suckling the mare for the first time, the foal should pass its first bowel motion. This is called meconium and is a thick, dark tar-like substance. If there is doubt about passage of meconium, lift the foal's tail. For some unknown but convenient reason, the passage of meconium nearly always leaves tell-tale evidence by way of a small dark wax-like blob on the anus or on the underside of the tail.

If meconium is retained the foal will be uncomfortable and will strain to no avail. Retention of meconium is an extremely common cause of death in the newly born foal. It begins a chain of:

- constipation,
- straining to point of exhaustion,
- reluctance to suckle properly,
- apathy,
- dehydration,
- shock and
- death.

Under natural conditions meconium retention is quite rare. Modern or artificial conditions seem to predispose foals to this type of constipation. The colt foal (because of its narrower pelvis) is affected much more frequently than the filly is.

However, forewarned is forearmed. There is seldom need to lose a foal from retention of meconium these days. Except in abnormal and very rare circumstances observation is the best preventative. Disposable, humans' enemas are cheap and readily available from chemists all over the world. So is paraffin oil.

Several weeks prior to foaling make sure you have three disposable enemas and 120 ml (about 4 fl oz) of paraffin in the store cupboard. It won't take much space if not needed and may well save a foal's life.

The constipated foal will give plenty of easy-to-diagnose signs that it needs attention. Continual or intermittent straining to no avail is characteristic. If the foal strains in this manner for longer than two hours during its first twenty-four hours an enema should be given according to the instructions on the package. Alternatively a 'pump' or 'can-type' enema can be used to administer approximately 300-600 ml (about 10-20 fl oz) of warm soapy water (mild soap please!) to the rectum. Glycerine may be substituted for part of the water if desired. The foal may be carefully given 60 ml (about 2 fl oz) of paraffin (orally) if somebody experienced can do it. Otherwise call a vet or other help.
NOTE: Watch for results after the enema; repeat if necessary.

We have, up until now, assumed that the foal suckled the mare soon after birth. Sometimes they don't! There can be many reasons for this; such as:
- the foal may be weak,
- it may have bent or otherwise distorted legs (which will prevent it standing), or
- the mare might be ticklish or cranky and not allow the foal to suck.

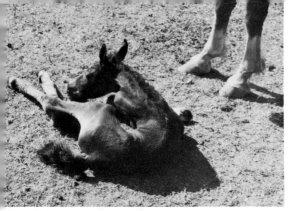

Curvature of the spine! This little foal couldn't unbend for several days. It was basin fed (on milk from its mother) until able to stand alone and suckle unassisted.

After several days the spine straightened and the foal was able to stand unassisted. It is shown here approximately one week old and is drinking from a basin.

It is well beyond the scope of this book to try to explain how to get a foal to suck a mare, or how to make the mare accept her foal when she doesn't want it. Both of these situations are abnormal and will need experienced assistance to rectify them. Until help arrives however the owner can easily help out in another way.

Almost without exception foals will, with a little perseverance, drink from a basin. It is of very great importance that the foal obtains colostrum within its first few hours of life. Colostrum is present in the first milk of a newly foaled mare and is an extremely concentrated form of nourishment for the foal. More important however is the fact that it contains antibodies (against infection) that are able to be absorbed through the lining of the foal's stomach (into its bloodstream) *only* during the early hours of the foal's life. Opinions vary on the length of time during which the colostrum may be so absorbed but all agree that it is well under thirty-six hours and possibly considerably less (perhaps as soon as six hours) in extreme cases.

Most studs now keep a bank of frozen colostrum so these days it is relatively easy to obtain some if the mare is unable to be milked.

In the immediate absence of colostrum the foal which hasn't suckled within a few hours of birth should be given some form of nourishment. Two easy formulas are:

1. Human baby milk substitutes (Lactogen, etc.) made up according to directions on the container for a three-month-old baby.
2. Evaporated milk, e.g. Carnation, Ideal, Bear) is available worldwide in corner shops, supermarkets, etc. and should always be kept on hand if a foal is expected. It is made up at the rate of one part evaporated milk to one part water. To each 240 ml (about 8 fl oz) of the made up mixture add one teaspoon golden syrup (or, in an emergency, sugar).

Feed either of the above mixtures at the rate of 240 ml (about 8 fl oz) each hour or hour and a half according to the foal's appetite. If fed hourly this will provide about 5 litres (10 pints) of milk in the first twenty-four hours. This is somewhere about the amount of milk the mare would produce during that time (production increases as the foal gets older).

Do not give larger quantities as this may upset the foal.

It goes without saying that these are only first aid measures and professional help (vet or studmaster) is usually needed in such a situation.

If the new-born foal shows any sign of physical abnormalities (i.e. twisted jaw, parrot mouth, etc.) call another breeder or your vet and seek initial advice. Seldom do these people mind genuine enquiries such as this and will do all they can to give assistance. It is important when enquiring about things which

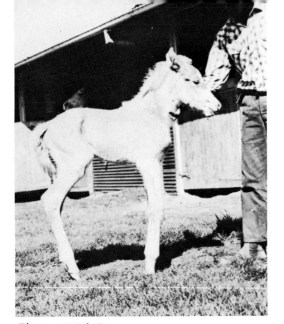

Photos: Keith Stevens
Above, right and far right are three examples of severe contracted tendons in day-old foals. All became 'normal' within two weeks of complete rest in a stable.

look abnormal (such as bent legs, contracted tendons, etc.) to get advice from people who see a lot of these deviations and therefore know from experience whether or not they are serious. With all due respect, it would be hard to find a small-animal practitioner who has had sufficient practical equine experience to give an opinion on these subjects.

Should the foal have any limb deviations it is advisable to lock the mare and foal up in a stable so it may rest. It may take several days (or weeks) but very rare indeed are the deviations that don't respond to this simple therapy which is so commonly overlooked.

Another common practice at foaling is to treat the foal's navel with iodine, antibiotics, methylene blue, etc. Although it is yet to be proven that it has any effect in the prevention of joint ill (navel ill) (which can actually enter the body other than through the navel) it doesn't do any harm!

A lot of large studs and a growing number of smaller ones are now finding it necessary (as a precaution against early bacterial infections) to inject each new foal with a high dose of a long-lasting antibiotic soon after birth. This practice has undoubtedly had a favourable impact towards prevention of bacterial scours in high-density breeding areas. Whether it is warranted under other circumstances is open to question.

Foal diseases and infections may be likened to measles, mumps, chicken pox, etc., in children—i.e. while on their own home ground in isolation children often remain unaffected but when they go to school all these diseases seem to attack them, one after another.

Scouring (severe diarrhoea) is another major cause of early foal death. When it occurs in a foal under thirty-six hours old it must be regarded seriously and help should be called immediately. Scouring should not be confused with the normal soft mustard-coloured motions which are passed after the meconium has been expelled. Severe scouring is sometimes overlooked in Arabian-bred foals because it is characteristic of these foals to normally carry their tails higher than other breeds. As a result the bowel motion squirts (yes, squirts) out below the upheld tail. In other breeds (when the tail is not carried so high) the liquid bowel motion will soil the tail from onset of scouring and is easily seen. (Soiled parts should be washed, dried

Photo: John Davis

A sick foal. Note the dried milk on its face below the eyes. Although only sick for some six hours, this foal has a listless appearance

and 'vaselined'.)

When scouring is severe the baby foal will stop suckling. It may appear to suckle but on closer observation will be seen to only 'mouth' the teat. The mare will quickly 'bag up' and milk will drip or spray from her teats on to the foal's face when it puts its head near the udder. If this occurs, call the vet immediately. (Home remedies for scouring foals are rarely beneficial in such severe cases.)

Again. . .it is important to wash the accumulated milk from the mare's hind legs.

Mild scouring of the foal often occurs during the mare's first (and occasionally second) heat period after foaling. This type of scouring is unlikely to be serious and seldom causes setbacks. Nevertheless the foal should be carefully observed and its tail and hindquarters kept clean.

Veterinary advice should be obtained soon after the birth of the foal to ascertain the most suitable age for its tetanus vaccination. This is essential. Immunization against strangles may also be advisable. Parasite control in the foal is an absolute must and a suitable programme should be chosen according to the conditions under which the foal is to be reared.

IN CONCLUSION

By the time a normal foal is four weeks of age it should have the strength, vigour and resistance to cope with most of the minor ailments that are likely to confront it before weaning. Nonetheless we must always remember and accept the fact that foals are animals, not humans, and no matter how much love and attention they are given we can never predict how they will react to any particular challenge; neither can we make them immune to or prevent all accidents and illnesses.

Breeders and owners who hope to rear healthy foals have a very real responsibility to carefully observe them as often as possible. At the first sign of anything abnormal or unusual a closer inspection should be made and the situation evaluated accordingly. Proper advice should be sought immediately and veterinary assistance secured if necessary.

If these basic guidelines are followed foal deaths will be kept to a minimum.

The ultimate aim of every breeder ... a healthy mare and a healthy foal.